The Garbage Menace

When we try to pick out anything by itself we find that it is bound fast by a thousand invisible cords that cannot be broken, to everything in the universe.

— John Muir

Dedication

This book is dedicated to Belizeans everywhere who genuinely care about their environment and the health of their fellow human beings. It is with hope that the contents of this book will inspire and encourage Belizeans to make wise choices about waste.

Footprint: A Belizean Environmental Series

The Garbage Menace is the first in this series aimed at enlightening Belizeans about the issues affecting their health and environment.

Other books planned for this series:

2—*The Mosquito Book*
3—*The Tainting of Paradise: Air Pollution in Belize*
4—*GMO or GM No?*
5—*Beyond the Cell Phone: Electromagnetic Radiation and the Invisible Threat*
6—*The Toxic Time-Bomb*
7—*The Energy Debacle*
8—*Intensive Farming and the Demand for the Fowl*
9—*When No More Is Left: Resource Depletion and Exploitation*

The Garbage Menace

Michael F. Somerville

Footprint: A Belizean Environmental Series 1

Published by *Producciones de la Hamaca*, Caye Caulker, Belize
<producciones-hamaca.com>

ISBN: 978-976-8142-764 (print edition)
ISBN: 978-976-8142-771 (e-book edition)
The Garbage Menace is the first in the series, *Footprint: A Belizean Environmental Series*
ISBN: 978-976-8142-788

Illustrations by Michael F. Somerville

Special thanks to the following sponsors for their contribution towards the development of this book: Belize Audubon Society, Belize Solid Waste Management Authority, Belize Tourism Board, Department of the Environment, Programme for Belize, and World Wildlife Fund Belize. The ideas and concepts in this book are solely those of the author and do not necessarily reflect the policies and objectives of the sponsors.

This book was printed on-demand by Lightning Source, Inc (LSI). The on-demand printing system is environmentally friendly because books are printed as needed, instead of in large numbers that might end up in someone's basement or a dump site. In addition, LSI is committed to using materials obtained by sustainable forestry practices. LSI is certified by Sustainable Forestry Initiative (SFI® Certificate Number: PwC-SFICOC-345 SFI-00980). The Sustainable Forestry Initiative is an independent, internationally recognized non-profit organization responsible for the SFI certification standard, the world's largest single forest certification standard. The SFI program is based on the premise that responsible environmental behavior and sound business decisions can co-exist to the benefit of communities, customers and the environment, today and for future generations <sfiprogram.org>.

Producciones de la Hamaca is dedicated to:
—Celebration and documentation of Earth and all her inhabitants,
—Restoration and conservation of Earth's natural resources,
—Creative expression of the sacredness of Earth and Spirit.

Contents

Foreword

Michael "Mike" Somerville's dream has come true with the publication of *The Garbage Menace*. An environmentalist by profession and from the heart, Mike's passion for the environment permeates every single page of this book.

The first and second parts of this book leave one with a sense of defeatism and helplessness. We seem to have a total disregard for the environmental state of our tiny nation, which is basically a dot on Planet Earth. The little dot that is Belize, however, is a giant among nations when it comes to the beauty of our natural and cultural environment—our people and cultures, our coral reefs and our forests. As Belizeans, we should be proud of what we have. Yet, we seem to take our country for granted. We indiscriminately dispose of garbage oblivious to the various waste categories and their hazardous components. We handle many items during the course of our daily lives unaware of the deleterious effects to our health and the environment. This information leaves you with a sense of wonderment: Does this concern me? Is there anything that we can do to reverse this trend? Should we even bother?

Thankfully, there is hope. In the final part of the book, Mike presents many things that Belizeans can do to get rid of the Garbage Menace. It is extremely promising to note that there are many good efforts that are taking place in Belize to properly manage waste. Many of the ideas presented can be put into action through our individual and collective efforts, and with reasonable resources. We can reverse the trend. We can defeat the Garbage Menace.

Mike's approach to this book is brilliant. It is easy reading and can be understood by academics and non-academics alike. It shakes you to your foundations at first, and leaves you wondering if all is lost. The book then ends with loads of ideas about how we can make our country, and the world, a better place to live.

The Garbage Menace is the first book in the *Footprint: A Belizean Environmental Series*, which is aimed at enlightening Belizeans about the issues affecting their health and environment. I cannot wait for the other books from the mind of my good friend and colleague, Michael "Mike" Somerville.

Osmany Salas
March 5, 2015

Preface

I got my initial inspiration to write an environmental book series after seeing a pool of used motor oil on a street in Ladyville. I realized then, as someone knowledgeable about environmental issues, that Belizeans need to be adequately aware of matters in their country that can affect their health and environment.

I recognize that the information on garbage in Belize, discussed in this first book in the series, can be obtained through browsing the internet, reading the newspaper, and listening to the television and radio; but this is the first time that a complete knowledge of the subject has been put together in one accessible and easy to understand source for Belizeans. This book is divided into three sections: "What Is Garbage, Anyway," "Hazardous, You Say," and "Is It Too Late for Anything to Be Done." The first section introduces garbage; presents a number of littering scenarios; provides possible explanations for why we litter; and gives an indication of how long it takes different things to decompose. The second section identifies the various waste categories and their hazardous components, and the effects of these hazardous materials on our health and environment. The third section informs about the many wise efforts that are taking place in Belize to properly manage waste, and things that we Belizeans can do to help this process. A glossary is provided to further explain unfamiliar words and terms used, and a reference list of the main sources is included for readers who may wish to further investigate some of the topics covered.

In writing this first book, I wish to thank the following individuals who, without their inspiration and support this book would not have been written: My heartfelt gratitude to my wife, Judith, for her support of this idea and for providing a fresh pair of eyes on the manuscript. Thanks also to my colleague, Osmany Salas, for his initial encouragement for me to write this book after seeing the first draft of the manuscript, and for steering me in the right direction. To my editors, Judy Lumb and Dorothy Beveridge, for their enthusiasm for my ideas, patience, commitment to the series, and very perceptive editing, you have my gratitude. I would also like to thank those individuals who took time out of their busy schedules to review the manuscript and provided their professional feedback: Dianne Lindo, Martin Alegria of the Department of the Environment, and Lumen Cayetano of the Solid Waste Management Authority. And lastly, to the many inspiring Belizeans, financial supporters, and other individuals from around the world whose work and dedication to protecting the environment and desire for a healthier world for all of us to live in, I thank you.

Michael F. Somerville
March 3, 2015

What is Garbage, Anyway?

Garbage, junk, litter, rubbish, trash, scrap, refuse, dirt, waste

Whatever we call it, it is usually something that:

- has reached the end of its life and is no longer useful,

- is no longer wanted,

- is no longer needed,

- is the by-product or residue of a process, or

- occurs as a result of activities by individuals or a community.

According to a Waste Generation and Composition Study done in 2010/2011 for the Western Corridor (Belize City, San Ignacio/Santa Elena, San Pedro and Caye Caulker) (population 100,000), the average Belizean throws away about two pounds of waste each day, without giving much thought to where it ends up.

But today, we should be mindful that the garbage we throw away can harm our health, the environment, and impact the world around us.

The Menace is Among Us

Have you ever noticed someone disposing of their garbage? Were you surprised how it was done? Some of these ways may or may not be familiar to you, but could easily be happening somewhere in Belize.

We have all at one time or the other handled our waste irresponsibly.

- A sign warns about the penalty for illegally dumping waste—at the foot of the sign is a large mound of garbage.

- In a village, a pile of garbage is burned in a back yard—and the smoke drifts through the homes of some residents further downwind.

- Scraps of lumber treated with wood preservatives are burned in a fire hearth.

- The driver of a sewage truck, under the cover of darkness, empties the truck's contents into a ditch.

- After changing the motor oil in a car, a mechanic pours the waste oil into the storm drain in front of his shop.

- The driver of a pickup truck stops along the side of a highway at sundown, then looks both ways before leaving an old refrigerator and scrap tires in a ditch.

- A passenger bus goes by—someone throws a bag of garbage through a window.

- On a trail, wild animals are attracted to the garbage left behind by hikers.

- Students play with syringes, needles, vials of liquid, and IV tubes discarded on a street near a school in a village.

- An old, broken-down passenger bus is left to rot and decay in its owner's back yard.

- A cruise ship with 3,000 passengers on board, releases its solid and liquid waste into Belize's territorial waters.

- After cutting and raking the yard, a neighbor throws the yard waste over her fence, into an adjacent lot that's overgrown and vacant.

- A garbage truck goes by in a hurry, leaving trash behind.

- A hospital burns human body parts in an open garbage dump—after its incinerator broke down.

- In an area of Belize City, residents use garbage as landfilling for their swampy property.

Why We Do the Things We Do

There are many reasons why someone would leave an old car to rot in his yard, throw garbage out the window of a bus, or leave an old refrigerator on the side of a road.

- It is easier and more convenient to throw garbage on the ground than to carry it some distance to a garbage bin.

- Garbage pickup service might be unavailable or inconsistent where we live.

- We have been littering for so long that it has become a habit.

- We might be unwilling to pay, or not able to pay to have our garbage hauled away.

- Some industries and businesses that generate certain types of waste might try to avoid having to pay the cost of disposal.

- The laws governing proper disposal of waste might be poorly understood, ineffective, or not adequately enforced.

- With an increasing population, more and more of us are generating waste.

- Some of us are wealthier than others and so are able to buy more things and have a lot more waste.

- Waste might fall out of a collection vehicle because it was not properly secured.

- Seeing litter around might make us want to litter as well.

- Some of us do not care about our health, the health of others, or how our surroundings look.

- Some of us do not know that waste can damage our health and the environment.

- Some of us might not be aware of the potential opportunities and benefits of reducing our waste or not littering.

- Some of us think that there won't be any consequences for our actions, or that when we litter there will always be somebody else to clean it up.

A Very Long Time To Decompose

- A banana peel may take several weeks to decompose!

- Paper takes several months!

- A tin can may take up to 100 years!

- Over 500 years for a plastic drinking water bottle and disposable diaper!

- A car tire may take up to 2,000 years!

- A million years for a glass soft drink bottle!

- And even longer than a million years for some kinds of plastic, like Styrofoam!

Many of us might think that the wastes we throw away will quickly rot, disappear into the ground, or be washed away.

In many cases this is not so. Some wastes break down very slowly and remain in the environment for an extremely long time!

Hazardous, You Say?

Pharmaceuticals and Personal Care Products

These include medicines, drugs, medications, cosmetics, fragrances, lotions, shampoos, soaps, tooth paste, nail polish, hair sprays, and hair straighteners.

Hazardous Materials: They contain parabens, phthalates, triclosan, lead, and volatile organic compounds (formaldehyde, glycol ethers, and acetone).

Health and Environmental Effects: These hazardous materials can affect our brain, heart, kidneys, liver, and reproductive and immune systems; cause diabetes, breast cancer and other cancers; pollute the environment; and kill wildlife.

Medical Waste

These includes syringes, blood, body parts, needles, gloves, scalpels, and some medical equipment from hospitals, health clinics, nursing homes, medical labs, and funeral homes.

Hazardous Materials: They contain radioactive and infectious materials.

Health and Environmental Effects: These materials can spread diseases (such as AIDS, hepatitis, and tuberculosis), and contaminate water and food sources.

The improper burning of these wastes can also release substances like polycyclic aromatic hydrocabons, dioxins, and furans into the air that can damage our brain, and cause birth defects and cancers.

Fluorescent Bulbs

They light up our home and workplace.

Hazardous Materials: They contain mercury. Broken bulbs can release mercury into the air and expose us to mercury from inside the glass and from the phosphor powder coating on the inside of the bulb.

Health and Environmental Effects: Mercury can cause blindness, damage our brain, kidneys, and lungs. Because it accumulates in fish (baca, botasi, grouper, tuna, snapper, bay snook, lobster, crab, tarpon, mackerel, tilapia, catfish, shrimp, sharks, and sardines), mercury is passed on to us when we eat these fish.

Chemical Pesticides

These include methyl parathion, Gramoxone, Roundup, Shelltox, Baygon, Klerat, and DDT, all of which we use for killing weeds, insects, and rats.

Hazardous Materials: Pesticides contain organophosphates, carbamates, organochlorines, volatile organic compounds, heavy metals, and other toxic substances.

Health and Environmental Effects: These toxic substances can affect our brain and kidneys, and cause Parkinson's disease, birth defects, and cancers. Some pesticides are not specific to particular pests. They unintentionally kill many useful animals like butterflies and bees that come in contact with pollen and nectar in the flowers of poisoned plants.

Chemical Fertilizers

Hazardous Materials: Fertilizers contain nitrogen compounds, persistent organic pollutants (dioxins, polychlorinated dibenzo-p-dioxins, and polychlorinated dibenzofurans), radioactive elements (uranium and polonium-210), and heavy metals (cadmium, uranium, mercury, lead, and arsenic).

Health and Environmental Effects: These chemicals can cause birth defects, lung cancer and other cancers; affect the brain, kidneys, and heart; create dead zones in the oceans; and kill wildlife.

Paints, Varnishes, Protective Coatings, and Glues

Hazardous Materials: Contain volatile organic compounds (benzene, methylene chloride, glycol ethers, and formaldehyde) and toxic heavy metals (lead, chromium, and cadmium).

Health and Environmental Effects: These chemicals can damage our brain, liver, kidneys and reproductive system; and cause cancers.

Household Cleaners

These include laundry and dishwashing detergents, furniture polish, glass cleaners, bleaches, soaps, disinfectants, oven cleaner, deodorizers and air fresheners, and bathroom and drain cleaners.

Hazardous Materials: These cleaners contain cyanide, chlorine, toxic heavy metals (mercury, chromium, and cadmium), and volatile organic compounds (formaldehyde, benzene, and chloroform).

Health and Environmental Effects: These chemicals can damage our liver, kidneys, and brain; and cause breast and lung cancer.

Refrigerants

They are used in air conditioners, refrigerators, and foam insulation.

Hazardous Materials: Refrigerants contain chlorofluorocarbons (CFCs), hydrochlorofluorocarbons (HCFCs), and hydrofluorocarbons (HFCs).

Health and Environmental Effects: These chemicals deplete the ozone layer and contribute to global climate change. HFCs do not damage the ozone layer but are global greenhouse gases that contribute to heating up the planet.

Engine Coolant or Antifreeze

Hazardous Materials: Antifreeze contains ethylene glycol and methanol.

Health and Environmental Effects: These chemicals can cause blindness; affect our brain, heart, lungs, and kidneys; and cause abnormalities in some bird species.

Tires

Hazardous Materials: Tires contain heavy metals, hydrocarbons, and other toxins that may leach into surface and groundwater over time; and release carbon monoxide, sulfur dioxide, butadiene, polycyclic aromatic hydrocabons, dioxins, furans, and styrene as air pollution when burned.

Health and Environmental Effects: These chemicals can damage our brain, liver, reproductive system and immune system; and cause heart and lung disease, and cancer.

Waste Oil

This includes transmission oil, brake fluid, and motor oil.

Hazardous Materials: Waste oil contains heavy metals (arsenic, lead, cadmium, and chromium) and other substances (benzene and polychlorinated biphenyls).

Health and Environmental Effects: These chemicals can damage our brain, heart, and kidneys; and cause birth defects and cancers.

Lead-Acid Batteries

They are used in cars, trucks, boats, motor cycles, and golf carts.

Hazardous Materials: Batteries contain sulfuric acid and lead.

Health and Environmental Effects: These chemicals can damage our brain, kidneys, and reproductive system; and severely burn our skin, causing blindness if they get in our eyes; and severely damage our internal organs causing death if swallowed.

Other Batteries

They include cell phones, watches, clocks, flashlights, radios, and small electronic devices.

Hazardous Materials: These batteries contain heavy metals (lead, mercury, and cadmium), and chemicals.

Health and Environmental Effects: The heavy metals and chemicals can cause brain and kidney damage, and cancers. Burning batteries can also cause them to explode and release toxic fumes.

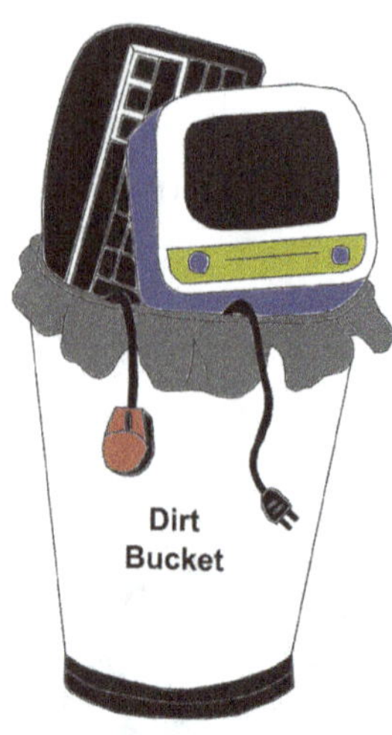

Dirt Bucket

Electronic Waste

These include discarded computers, microwaves, smoke alarms, cell phones, printers, copiers, television sets, cameras, and radios.

Hazardous Materials: These products contain heavy metals (mercury, cadmium, chromium, and lead), radioactive materials, hydrocarbons, and flame retardants.

Health and Environmental Effects: These chemicals can cause brain and liver damage, thyroid problems, cancers, and harm the environment. Open burning of electronic waste also releases dioxins, furans, and other harmful substances into the atmosphere that can damage our brain and cause cancers.

Aerosol Cans, and Propane and Butane Gas Tanks

They are pressurized and can explode if punctured or burned, resulting in serious injuries.

Wastes from Industries

Manufacturing, farming, sewage systems, construction, demolition, mining, dry cleaning, school and research laboratories all produce hazardous waste.

Hazardous Materials: This waste contains heavy metals, chloroform, radiation, pathogens, asbestos, perchloroethylene, and other toxic compounds.

Health and Environmental Effects: These chemicals can damage our liver and kidneys; and cause cancers and birth defects.

Household Furniture

This includes bedding, carpeting, foam mats, sofas, and baby strollers.

Hazardous Materials:
Furniture contains flame retardants to make them fire resistant and fire proof.

Health and Environmental Effects: Flame retardants can cause infertility and birth defects; and liver, kidney, testicular, and breast cancers.

When household furniture and items treated with flame retardant chemicals are burned, the chemicals may react with other toxic materials in the smoke to produce cancer-causing dioxins and furans.

Plastics

Hazardous Materials: Plastics contain styrene, pthalates, adipates, alkylphenols, vinyl chloride, and bisphenol-A.

Health and Environmental Effects: These chemicals can affect our reproductive and immune systems, damage our brain and liver, and cause heart disease and cancers.

Open burning of plastics releases heavy metals, polycyclic aromatic hydrocabons, dioxins and furans, carbon monoxide, benzene, and lots of carbon into the atmosphere that contribute to global climate change.

Improperly disposed of plastics (cigarette butts, plastic bags, fishing lines and nets) can also clog drains, and get into rivers, the sea, and other water bodies, where they kill fish, sea turtles, and seabirds that become entangled in them or mistakenly feed on them.

Waste can affect us in other ways, too!

- Littered garbage can cause visual pollution, which might be a source of public nuisance, stress, anxiety, and even a distraction to some people.

- Litter falling from transport vehicles can injure pedestrians and cause automobile accidents.

- The high cost to clean up improperly and illegally dumped garbage might cause the local government to have to divert funds from other important programs like education, healthcare, and infrastructure.

Is It Too Late for Anything to Be Done?

Despite the size of the garbage problem, there is still hope. Today, it may be necessary to combine several methods to manage our waste. There is no single, simple solution to our garbage problem. Some wise efforts are underway in our country, and others we can make to ensure our waste does not menace us into the future.

We Can Contain Our Waste Better

A good start to taking good care of our waste is to make sure that it is properly separated and contained.

Our communities can do with proper waste storage containers. Waste containers can be made available to households and/or through a community waste storage and collection system. Containers can be placed in common and convenient locations to discourage littering. They can be maintained and emptied in a timely and efficient manner to prevent overfilling and unsightly searches that can add to the already-existing litter problem.

Belize is making serious efforts to properly contain waste by constructing Waste Transfer Stations in the Western Corridor, where waste from residents in this region is collected and then transferred to the Regional Sanitary Landfill at Mile 24 on the George Price Highway. Similar facilities are envisioned to address the waste disposal needs of the remainder of the country.

We Can Practice Recycling

A lot of the waste that we throw away can be recycled into similar or new products, and the more we recycle, the less waste will go to our open public dumps, our landfill, and to polluting our environment.

Recycling also saves money and energy, and spares our natural resources by reducing the need to extract more raw materials to make new products.

There are many recycling companies in Belize that offer drop-off locations or collection centers where we can take our recyclables. Some of these companies may even pay us or offer us some other incentives for our recyclables.

Recycle Plastics

Recyclable plastics can be melted down and reformed into clothes, plastic lumber, tables, chairs, and trash cans.

Almost all plastics have a number inside a triangle of arrows marked on them. This helps to identify the types of plastic for recycling.

Number 1 and 2 plastics are the most commonly recycled plastics in the world today and may be the only ones accepted by the recycling companies in Belize.

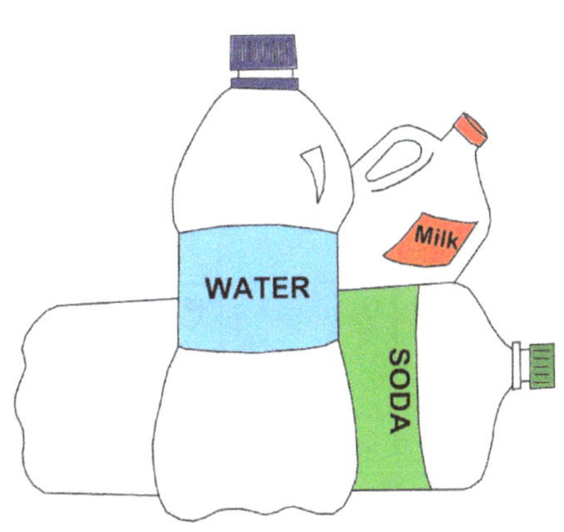

Some companies and initiatives that recycle plastics in Belize include, Bowen and Bowen Limited (shreds and re-molds its bottles), Belize Recycling Company (shreds some bottles), and Resource Recovery Recycling Limited.

In addition, we can voluntarily drop-off our recyclable plastics at a waste transfer station in our area.

Types of Plastic

Number	Plastic Type	Where Found
1	PET	Soft drink bottles Water bottles
2	HDPE	Grocery Bags Milk jugs Laundry detergent bottles Plastic lumber Ice cream containers
3	PVC	Shampoo bottles Shower curtains Lawn chairs Toys Water pipes
4	LDPE	Plastic bags Squeezable bottles Clothing Furniture Carpet
5	PP	Syrup bottles Straws Food containers Dishware Auto parts
6	PS	Meat trays Egg cartons Compact disc cases Food containers Toys Utensils
7	Other	5-gallon water bottles Sunglasses Car headlight lenses Safety shields

PET = Polyethylene Terephthalate
HDPE = High-Density Polyethylene
PVC = Polyvinyl Chloride
LDPE = Low-Density Polyethylene
PP = Polypropylene
PS = Polystyrene

DID YOU KNOW?

EcoFriendly Solutions Limited is a company in Belize that sells plastic replacement products (such as plates, cups, forks, spoons, knives, trash bags and shopping bags) made from natural materials.

DID YOU KNOW?

Many countries such as South Africa, China, Taiwan, Bangladesh, and Macedonia already have a total ban on light weight plastic bags.

Recycle Metals

Copper, aluminum, bronze, stainless steel, brass, steel, tin, and iron are recyclable.

Copper can be found in electrical wires, air conditioners, and gas-line tubing for stoves.

Aluminum is usually used to make soda cans, gutters, sidings, and window frames.

Brass can be found in keys, door handles, and in light, bathroom, and plumbing fixtures.

Steel and iron can be found in cars, railings, fences, and bed frames.

Scrap metal recyclers, including the Belize Recycling Company, Belize Metal Recyclers, Orange Walk Metal Recyclers, Trans Metal, and Southern Metal Recyclers, may purchase certain scrap metals from a seller and export to countries such as Mexico, Guatemala, Taiwan, India, and China.

We can voluntarily drop-off our recyclable metals at a waste transfer station in our area.

Recycle Paper

Many products can be made from recycled paper, such as, writing paper, paper towels, paper money, lamp shades, toilet paper, egg cartons, and bandages.

Paper recycling companies in Belize (such as the Belize Recycling Company and Resource Recovery Recycling Limited) may take paper such as newspapers, telephone directories, office paper, magazines, exercise books, and cardboard.

We can also voluntarily drop-off our recyclable paper at a waste transfer station in our area.

Recycle Glass

Glass can be recycled into new glass or building materials like concrete, insulation products, sand paper, and water filters.

Colorless, green, brown or amber are the most commonly recycled glass.

Many glass recyclers in Belize accept glass like those used as containers for beverages, such as, soft drinks and beer. Businesses throughout Belize that sell these products offer a refund to customers for the return of the empty bottles.

Some companies may buy glass and export to countries such as Mexico and Guatemala where the technology is available for proper recycling.

Recycle E-Waste

Recycling electronic waste may be the most effective solution to the growing e-waste problem.

Most electronic devices contain a variety of materials, including rare earth metals that can be recovered and used again.

Resource Recovery Recycling Limited accepts all kinds of e-waste at its recycling plant in Ladyville.

We can also voluntarily drop-off our e-waste at a waste transfer station in our area.

Recycle Lead-Acid Batteries

The lead in lead-acid-batteries can be recycled into new batteries or new products.

Some businesses that sell lead-acid batteries may collect used batteries for recycling, and RENCO Battery Factory will pay us for our old battery when we purchase a new one.

Resource Recovery Recycling Limited also accepts these batteries at its recycling plant in Ladyville.

Some used batteries are re-built here in Belize but the majority are exported to countries such as the USA, China, and El Salvador where they are recycled into new batteries or products.

DID YOU KNOW?

From 2011 to 2014, Belize imported about 126,471 lead-acid batteries.

Recycle Other Batteries

Button-cell batteries and other batteries such as alkaline, nickel–cadmium (NiCd), nickel metal hydride (NiMH), and lithium-ion (Li-ion) can also be recycled.

Battery drop-off point or collection centers may also be available elsewhere in the country.

A good practice would be to dispose of these batteries along with our e-waste.

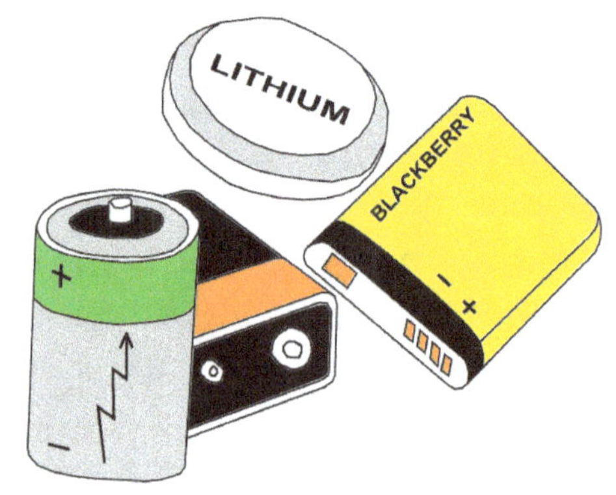

Recycle Tires

Tires can be recycled to make other tires, as well as products like shoes, rubber tiles, and floor mats.

Old tires can be converted into play swings, tire art, and pots for yard plants.

Used tires can also be retreaded or recapped instead of buying new ones.

Belize currently does not have a modern facility that processes used tires into new tires or products, but Mike Reb Automotive Group in Orange Walk Town does automotive tire recapping.

Recycle Waste Oil

Waste oil such as used motor oil can be recycled or refined into new motor oil, processed into fuel oils, and used as raw materials for the petroleum industry.

A lot of waste oil generated in Belize is not disposed of properly.

Belize currently does not have a national used oil recycling program, but some companies in Belize collect used oil for export to countries like Guatemala and Mexico, where it is converted to fuel.

Recycle Ink And Toner Cartridges

Ink and toner cartridges used in printers and photocopiers can be refilled, refurbished, or recycled.

Belize currently does not have a national ink and toner cartridge recycling program, but the Cartridge Recycle Company refurbishes, refills, and buys used cartridges, and the Copier Engineering Services re-manufactures toner cartridges.

Recycle Paints

Leftover paints can be recycled into other usable paints, or used in the manufacture of cement or other products.

Some hardware stores that sell paint may collect leftover paint for recycling.

Belize currently does not have a paint recycling program.

Recycle Fluorescent Light Bulbs

Fluorescent light bulbs can be recycled into new products.

The glass tubing can be turned into new glass prod-ucts, the brass and aluminum in the end caps can be reused, the internal phosphor coating can be reprocessed for use in paint pigments, and the mercury contained in the lamp can be reclaimed and used in new lamps.

Fluorescent light bulbs can be sep-arated from our general waste for recycling or safe disposal.

Belize currently does not have a flu-orescent light bulb recycling program.

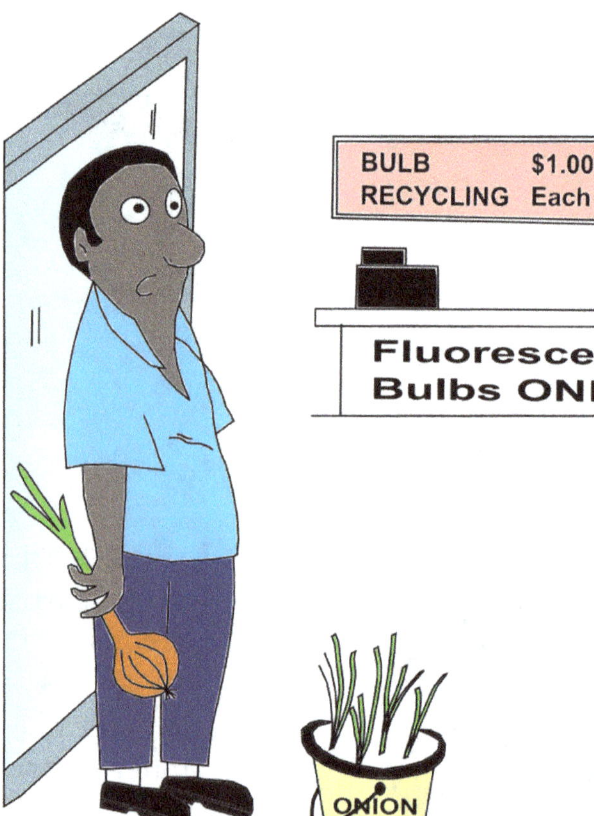

We Can Landfill Our Waste

Another way to deal with our waste is by using landfills.

A modern landfill is a managed and organized facility where waste is disposed of by burial and where harmful substances in the waste are prevented from contaminating the land, air, and water.

Putting our waste in landfills reduces the amount of harmful public dump sites.

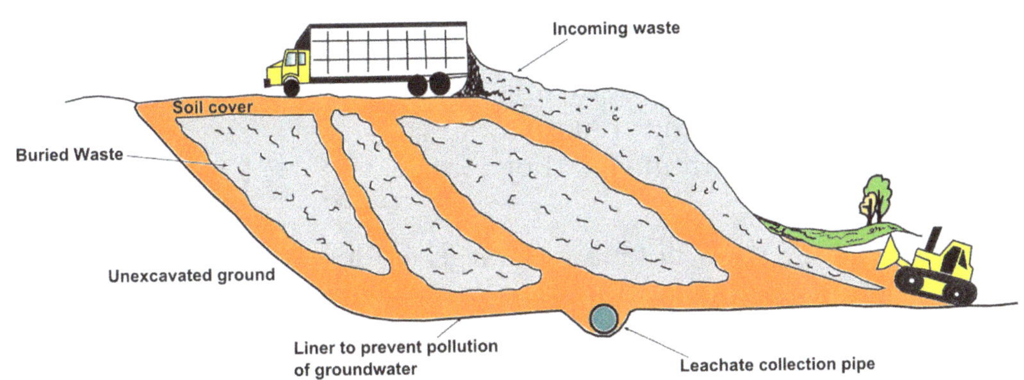

Hazardous Waste Landfill

Hazardous materials are usually not disposed of in a regular sanitary landfill, but in a separate and more securely managed facility called a "hazardous waste landfill" or "hazardous waste cell."

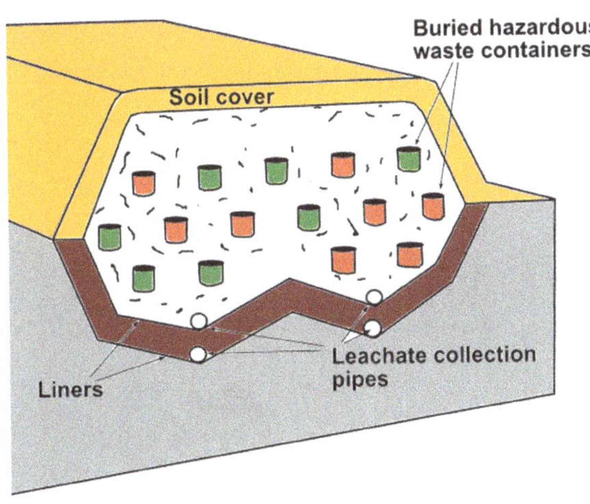

Belize has a modern hazardous waste cell at the new Mile 24 sanitary landfill facility that will be used for disposal and storage of pre-treated and containerized hazardous waste mainly from industries.

We Can Burn Our Waste in an Incinerator

A modern incinerator is a controlled and managed facility that uses high temperature burning to reduce the weight and volume of waste, while also preventing the harmful components in the exhaust from polluting the air.

Through incineration, less waste will go to open public dumps and landfills.

Some hazardous and medical wastes are best disposed of by this method to destroy the harmful pathogens and contaminants they contain.

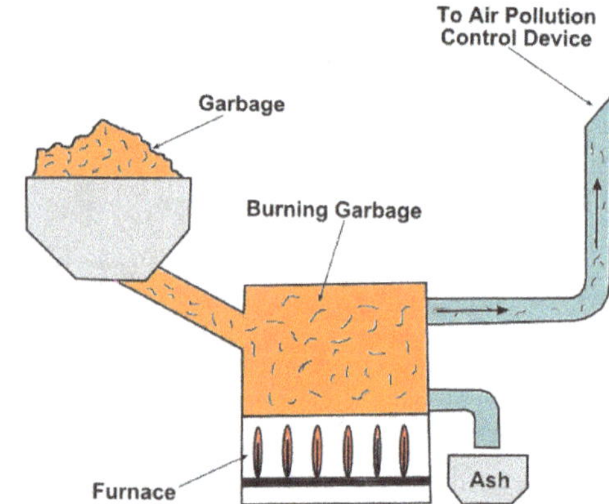

Belize currently does not have a large modern incinerator to burn household and medical waste, but the Belize Waste Control Ltd currently uses a small incinerator at its Mile 3 George Price Highway headquarters to burn medical waste from the Karl Heusner Memorial Hospital and other medical institutions in Belize City. Similar incinerators exist at some other medical establishments in the country.

Waste-To-Energy

Another benefit of incinerating waste is that it can be used to generate electricity.

The heat from the burning waste is used to boil water to produce high pressure steam that powers a steam generator that makes electricity.

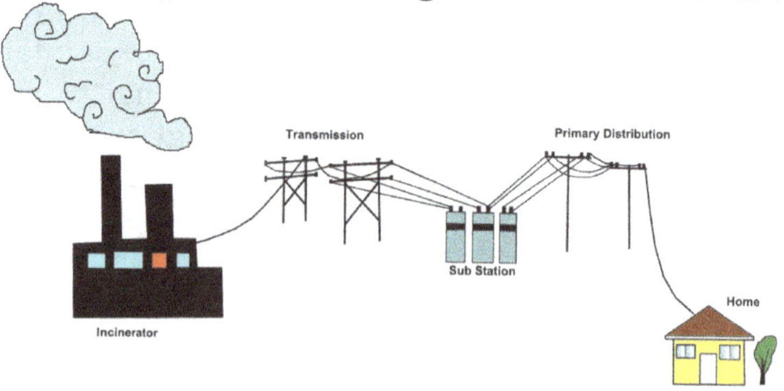

We Can Compost Our Waste

Materials such as yard cuttings and plant food scraps (banana peel, onion skin, plantain skin, potato skin, etc.) that are thrown in the garbage can be turned into compost.

Compost can be used as a natural fertilizer for plants and crops, and as a natural pesticide to control insects and weeds.

Composting with Aunt Grace

Mrs. Grace Grant, affectionately known as Aunt Grace, has been doing composting for over 50 years. It all began after she got married at the age of 38 when she and her husband used to visit a beautiful farm on the Northern Highway every Sunday in their little Jeep. Visiting the farm made her desire to have a little garden of her own where she could grow crops and beautiful trees. She has been composting ever since.

Today, she cannot commit as much time to composting and tending her garden like before but she still separates her fruit and vegetable skins from her regular garbage. She hates the thought of seeing potentially good natural fertilizer not being used and believes that Belizeans can save a lot by growing their own backyard garden and using compost.

Aunt Grace has been living at her Lindo's Alley address in Belize City since 1970, and will turn a vibrant 90 year-old in 2015. She is a practicing Bahai and credits her good age to the abundant exercise she got from composting and tending a garden, eating the right foods, and drinking bush tea (sage and lemon grass).

Build an Easy Compost Pile at Home

1) Make a compost bin about 3 feet long by 3 feet wide by 3 feet high out of wood or chicken wire.

2) Place plant food wastes and yard cuttings in the bin and keep adding to it.

3) Keep the pile moist but not soggy.

4) Turn the pile about every 2 weeks so air can flow through and help the waste to decompose.

5) Signs of steam rising from the pile and the presence of earthworms mean the composting process is working.

6) The compost is ready when it turns black and is crumbly to the touch.

7) Start a new pile with any leftover waste material that did not fully decompose.

COMPOSTING

The Belize Botanic Gardens in the Cayo District offers classes on composting as part of its Professional Gardener's Program.

Belize Has Good Laws

Belize Signed International Agreements

Because many of the influences and challenges contributing to our waste problems here in Belize are global and beyond our borders, we certainly cannot act alone, but must join with regional and international efforts to adequately address these challenges.

Convention for the Prevention of Pollution from Ships (Marpol 73/78) aims to prevent and minimize operational and accidental pollution (of garbage, oily wastes, sewage, etc.) from ships.

The Basel Convention aims to ensure the environmentally sound management of hazardous waste, as well as greater control of their movements between nations.

The Stockholm Convention aims to restrict or eliminate the production, use, release and storage of persistent organic pollutants or organic compounds that are resistant to decomposition and can greatly affect human health and the environment.

London Convention on the Prevention of Marine Pollution by Dumping of Wastes and Other Matter aims to protect the marine environment from the effects of dumping wastes and chemicals into the sea.

Montreal Protocol is designed to protect the ozone layer by phasing out the production and use of the substances that are responsible for its depletion.

Belize Laws Address Waste Issues

Solid Waste Management Authority Act empowers the Solid Waste Management Authority to manage the collection and disposal of solid waste in Belize.

Environmental Protection Act empowers the Department of the Environment to prevent, control and manage pollution, including the dumping or disposal of any wastes (including toxic and hazardous wastes) in any place in Belize, including the sea.

Public Health Act prohibits the illegal deposit or accumulation of any waste that is offensive, injurious or dangerous to health, or liable to favour the spread of any infectious disease.

City Council Act empowers a city council (or other urban authority) to coordinate, control, manage or regulate the timely and efficient collection and removal of all garbage material from all residential or commercial areas in a city.

Under the City Council Act, it is illegal for anyone to throw garbage onto the streets or in any improper place, accumulate garbage upon his premises, disturb garbage placed for collection and disposal, or place any potentially explosive material in a garbage collection bin.

In cities and towns throughout Belize, the local government collects and dispose of garbage from households free of charge. Only commercial users or businesses are required to pay to have their garbage hauled away.

Individuals or businesses who take their own garbage to a public disposal site are required to pay based on the amount that they throw away.

The Village Councils Act empowers a Village Council to make by-laws for the cleanliness of streets and other public places and for ensuring that sound environmental practices are adhered to by all persons within the village.

Environmental Tax Act provides for the collection of two percent tax on every good imported into Belize, to develop a national solid waste management program, to defray the cost of the disposal of refuse generated by the use of goods, to assist in the collection and disposal of garbage throughout Belize, to clean up rivers and canals and other internal waterways, for the preservation and enhancement of the environment, and for strengthening the institutional capacity of the Department of the Environment.

Port Authority Act empowers the Belize Port Authority to restrict and control the depositing of any substance, solid matter, article or thing that causes pollution of any port; and prohibits the depositing, placing, or discharging of any polluting substances into the territorial waters of Belize.

Belize Water Industry Act (formerly called Water and Sewerage Act) empowers the BWSL to prevent the pollution of any surface water or groundwater used for human consumption or domestic purposes.

Pesticides Control Act gives the Pesticides Control Board the power to control the manufacture, importation, sale, storage, use, and disposal of pesticides.

Returnable Containers Act authorizes businesses that sell beverages in glass or metal containers to collect a deposit and offer a refund to customers for the return of these containers.

Dangerous Goods Act makes provisions for the transport and safe storage of explosives and other dangerous goods.

Belize Could Improve upon Existing Laws and Make New Laws

Although our existing laws addressing our waste issues here in Belize are good ones, sometimes laws need to be improved upon and new ones need to be formed to adequately deal with current or new situations. Needed new laws would:

- ensure that businesses take responsibility for the recycling of the products they import and/or sell,

- revise the Returnable Containers Act to include non-beverage or plastic containers,

- ban, phase-out, or restrict use of light weight plastic bags, or require businesses to charge customers for plastic bags,

- ensure that public establishments put a proper waste management system in place,

- ban or restrict use and distribution of certain harmful plastic containers, especially those used for carrying or packaging food, such as styrofoam,

- ban paper receipts containing bisphenol-A and encourage merchants to use an electronic receipts system,

- ensure proper recycling and disposal of waste tires,

- regulate proper management and disposal of medical waste,

- regulate proper disposal of electronic waste and monitor the end-of-life treatment of electronics, and

- regulate proper disposal of household hazardous waste.

Belize Could Also...

- continue to run nationwide public awareness campaigns about good waste practices to reach all Belizeans;

- put convenient and easily-accessible facilities in place to encourage Belizeans to drop off recyclable materials separated from their household waste;

- seek out new opportunities to initiate or improve the recycling of plastics, paper, metals, and glass;

- create the enabling environment that will incentivize the private sector to want to recycle and reduce the amount of waste they generate;

- create or strengthen the local markets for recyclable materials;

- institute a composting system for villages and municipalities, and distribute the nutrient-rich compost fertilizer to urban gardeners and local farmers

- seek out new opportunities to improve the waste oil collection and recycling program;

- institute a used lead-acid battery collection and recycling program, including compelling all businesses that sell new or refurbished batteries to collect a refundable deposit from buyers;

- institute a program for the collection and proper recycling of fluorescent light bulbs, leftover paints, and other batteries;

- better enforce its existing environmental and waste disposal laws;

- ensure that all waste generated in the country undergo pre-treatment before final disposal at a sanitary landfill;

- ensure that its waste management facilities and systems are in accordance with international standards;

- build the capacity of waste management workers through adequate training and compensation;

- offer tax breaks to companies that manufacture and sell environmentally friendly products; and

- monitor the entire lifecycle of dangerous pesticides and other chemicals imported into Belize.

As Belizean Citizens, We Can...

Reduce

One of the simplest solutions to dealing with our waste problem is to produce less waste in the first place.

- Many of the things we buy we do not need, so we can reduce the amount of waste we generate by buying less and using less.

- We can rent or borrow items we do not use often.

- We can buy groceries and other products in bulk to avoid having to dispose of excessive packaging.

- We can use a water tank to collect rainwater for drinking and cooking, thereby reducing our dependency on purchasing water in plastic bottles.

- We can make our next event a zero-waste one by reusing event materials and setting up a zero-waste station for those recyclable and compostable items such as paper cups and plates, food scraps, and plastic water bottles.

- We can refuse cash register receipts that may contain bisphenol-A when making purchases or visiting the ATM.

Reuse, Repair or Refurbish

We throw away a lot of things that are reusable and could be useful to others.

We can:

- Shop from thrift stores and donate used clothes and household items to organizations like the Salvation Army and Humana that help the elderly and needy individuals.
- Organize garage sales and sell our used items instead of throwing them away and adding to the garbage.
- Reuse as many of our household items as possible, such as containers for storing food.
- Purchase items that last longer and are easier to recycle, rather than cheap items that break after only a short time of use.
- Use cloth diapers or those made from natural materials instead of plastic ones.
- Open goodwill stores to accept donations of used clothing and household goods and then resell these at a cost that is affordable to the poor.
- Encourage the practice of hand-me-downs.
- Use reusable shopping bags and minimize the amount of light weight plastic bags we take home from the grocery stores.
- Consider repairing or refurbishing an item before purchasing a new one.
- Purchase rechargeable batteries instead of single-use ones, for devices that we use frequently.
- Use LED light bulbs in our homes, that save energy and contain no toxic materials.

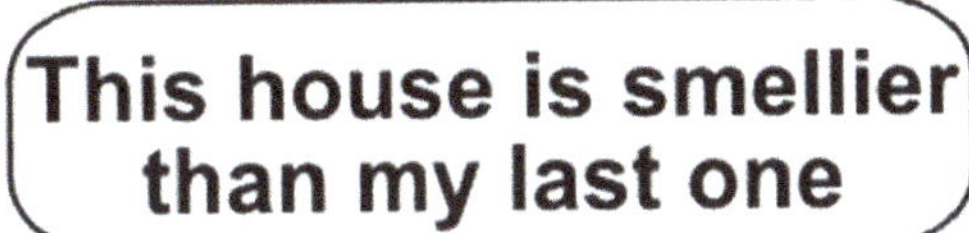

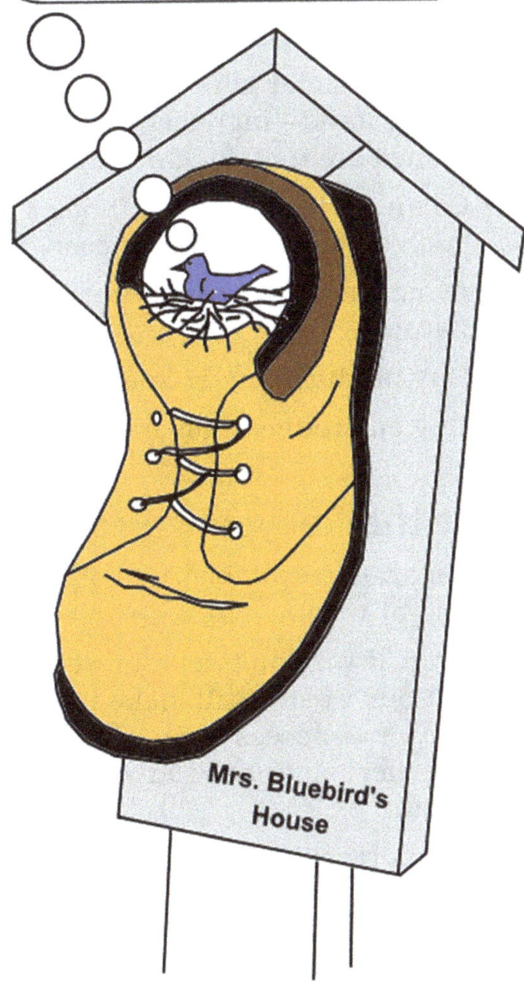

Buy and Use Products that are Safer for Our Health and the Environment

- Buy food and drink products contained in glass, metal (preferably stainless steel), wood or paper rather than plastic or styrofoam.

- Avoid paper plates that are coated with harmful leak-proof non-stick chemicals.

- Buy furniture and other items for our homes that are made free of flame retardant chemicals, and also avoid those treated with non-stick chemicals to repel stains.

- Buy organic or naturally grown foods or food products made from natural ingredients.

- Buy personal care products that list ingredients that we can understand—ingredients with long, difficult-to-pronounce names are usually harmful.

- Clean and freshen our homes with natural and safe products such as baking soda, vinegar, and lemon juice.

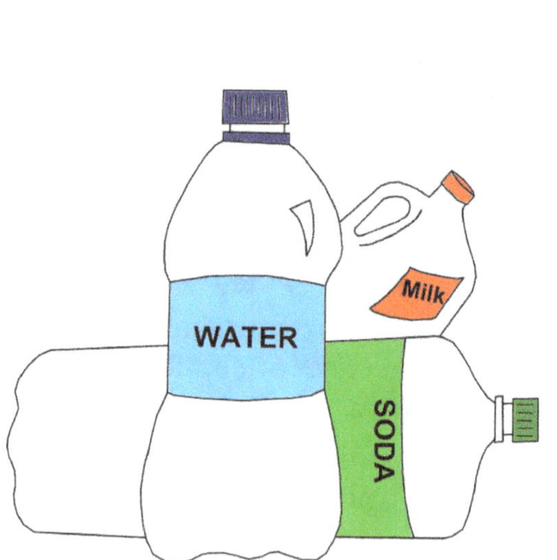

- Avoid any kind of air fresheners and deodorizers containing fragrances that are not natural.

- Buy products made from recycled materials or natural materials that are biodegradable.

- Buy environmentally friendly paints and donate leftovers.

Start the Recycling Process

- Separate recyclables and other wastes from home and at work.

- Even if we don't sell our recyclables, separating out our wastes will make it easier for people who recycle to access them when they reach the waste transfer stations and public garbage disposal sites.

- Support recycling projects in our communities.

- Share our recycling experience with others.

Use Safer Alternatives to Harmful Pesticides

- Use pesticides made from natural sources to treat insect pests on our farms, garden plots, and in and around our homes.

- Practice crop rotation to keep pests and plant diseases to a minimum by growing different crops on our plots of land for each new growing season.

- Grow plants (such as geranium, basil, lemongrass, chrysanthemums, marigold, citronella, catnip, peppermint, rosemary, thyme, and sage) alongside our crops and around our houses to naturally repel insects.

Use Organic instead of Man-Made Fertilizers

Manure from farm animals, wood chips, sawdust, and compost make good natural fertilizers.

Give Support

- Participate in garbage cleanup events in our neighborhoods or communities.
- Volunteer with groups or organizations cleaning up stretches of beaches, roads or highways.
- Join volunteer organizations that encourage healthy living and promote awareness of environmental and conservation issues.
- Support ideas and businesses that promote and sell safer products.
- Support and vote for political candidates that care about human health and the environment.

- Lobby for and support laws that encourage occupational health and safety.
- Participate in Go Green initiatives in our communities or workplaces.
- Help our municipalities to keep our communities clean and healthy by showing our willingness to pay garbage collection fees.
- Give greater recognition and value to sanitation workers who collect and transport waste.

Act Responsibly

- Avoid injury to sanitation workers and help our local authorities to better collect and dispose of our garbage by properly preparing and containing our waste at home and work.
- Store chemicals properly and in a safe location at our homes and workplaces to avoid contaminating ourselves and the environment.
- Become informed about environmental matters in our country.
- Conserve and avoid wasteful and extravagant use of resources.
- Report reckless and suspicious waste disposal activities to the authorities.

Afterword

The proper disposal of our waste is a responsibility for each of us. Today, our health and environment are showing signs of stress under our garbage footprint. Like never before, we are being confronted by the threat of the rise in cancers, heart disease, diabetes, and communicable diseases; our air, land and waters are being polluted; and large numbers of our country's wildlife are under threat of disappearing. But whether our garbage footprint is eventually altogether lessened, or becomes an even greater menace to Belize, will ultimately depend on the extent of care, civic pride, and understanding we commit as individuals and as a nation.

Glossary

Acetone: A colorless liquid that evaporates, burns easily, and is mainly used to dissolve or loosen another substance.

Aerosol: A mixture of fine solid or liquid particles in a gas.

Adipates: Substances added to plastics to make them softer and more flexible.

Alkylphenols: Substances added to many different kinds of plastic to make them more stable and usable.

Arsenic: A naturally-occurring heavy metal used in pesticides, paints, and for preserving wood.

Asbestos: A mineral fiber found in rocks and soil, used widely in the automotive and construction industries.

Baygon: A substance used for killing and controlling insects.

Benzene: An industrial chemical found in crude oil and gasoline, and used to make products including plastics, lubricants, dyes, detergents, and drugs.

Biodegradable: Capable of being broken down by microorganisms.

Bisphenol-A (BPA): An industrial chemical used primarily to make plastics.

Butadiene: An industrial chemical most commonly used to manufacture synthetic rubber as used in automobile tires.

Cadmium: A naturally-occurring heavy metal used mainly in the production of batteries, pigments, metal coatings, and plastics.

Carbamate: A chemical substance used in the production of pesticides to kill insects.

Carbon Monoxide: A colorless, odorless, and tasteless gas that results from the burning of natural gas and other materials containing carbon, such as, gasoline, kerosene, and wood.

Chlorine: A chemical gas or liquid with a suffocating odor, primarily used for bleaching, disinfecting, and purification.

Chlorofluorocarbons: Chemical compounds containing chlorine, fluorine, and carbon, used in the production of aerosol sprays, foam insulation, solvents, and refrigerants in refrigerators, freezers, and air conditioning systems.

Chloroform: A colorless, sweet-smelling liquid once widely used as a cleaning agent and anesthetic during surgery.

Chromium: A naturally-occurring heavy metal used for chrome plating, dyes and paints, tanning leather, and preserving wood.

Cyanide: A very poisonous chemical substance.

DDT: A chemical substance once widely used to kill insects in agriculture and insects that carry diseases.

Dioxins: Chemical compounds produced naturally and as a by-product of industrial activities, such as, pesticide manufacturing and burning of waste—known to linger for a long time in the environment.

Electronic Waste: Electrical or electronic devices that are no longer wanted or needed.

Ethylene Glycol: A clear, odorless, sweet-tasting chemical found in many household products, including antifreeze, detergents, paints, and cosmetics.

Flame Retardants: Chemicals added to manufactured materials such as plastics, textiles, and furniture to make them less likely to burn and spread a fire.

Fluorescent Light Bulb: A light bulb that contains mercury vapor and lined with a phosphor coating on the inside.

Formaldehyde: A strong-smelling, colorless, flammable gas or liquid primarily used to preserve human and other organic remains.

Furans: Toxic chemical substances produced during the manufacture of other chemicals or products.

Glycol Ethers: A class of chemical solvents widely used to manufacture cleaning products and protective coatings.

Gramoxone: A chemical pesticide used for killing weeds.

Greenhouse Gas: A gaseous compound that traps and holds heat in the atmosphere.

Hazardous: Posing threat to life, health, property, or environment.

Heavy Metals: Metallic chemical elements that have a high density and are toxic or poisonous at low concentrations.

Hydrocarbons: A class of chemical compounds that consist entirely of the elements hydrogen and carbon, and found mainly in crude oil.

Hydrochlorofluorocarbons: Chemical compounds similar in structure to chlorofluorocarbons, including cholorine, fluorine, and carbon, but also hydrogen.

Hydrofluorocarbons: Chemical compounds containing hydrogen, fluorine, and carbon—used as replacement for chlorofluorocarbons and hydrochlorofluorocarbons.

Klerat: A chemical compound used to kill rats and mice.

Leachate: A contaminated liquid formed as water drains through garbage and picks up its harmful components.

Lead: A naturally-occurring heavy metal used in a wide variety of products, including paints, batteries, gasoline, cosmetics, and electronics.

Light Emitting Diode (LED): an electronic device that gives off light when an electrical current is passed through it, and is used in lighted signs, clocks, flashlights and light bulbs.

Menace: A person or thing that threatens to cause harm or injury.

Mercury: A naturally-occurring heavy metal used in the manufacture of industrial chemicals, electrical switches, thermometers, electronics, and light bulbs.

Methanol: A toxic, colorless chemical used as an industrial solvent, pesticide, and alternative source of fuel.

Methyl Parathion: A highly toxic chemical used to control and kill insects on agricultural crops.

Methylene Chloride: A colorless, sweet-smelling liquid that evaporates easily—primarily used as a paint stripper and remover, to clean and degrease metals, and in the manufacture of pharmaceuticals, glues, and foams.

Organic: a substance of, relating to, or derived from plant or animal.

Organochlorines: Chemical compounds containing carbon, chlorine, and hydrogen that are highly insoluble in water but are readily stored in fatty tissues of humans and animals.

Organophosphates: A class of highly toxic chemicals used to control and kill insects on agricultural crops.

Ozone Layer: A region of the Earth's atmosphere that protects living things from the sun's harmful radiation.

Parabens: A group of chemicals used to preserve pharmaceuticals and personal care products due to their abilities to kill bacteria and fungi.

Pathogen: A bacteria or virus that causes disease or illness.

Perchloroethylene: A sweet-smelling, colorless, non-flammable liquid—used mainly to dry clean fabrics, degrease metal parts, and as a paint stripper.

Persistent Organic Pollutants: Toxic chemicals that are transported around the world by wind and water, and are resistant to breakdown in the environment.

Pesticide: A chemical substance or mixture intended for controlling, repelling, or destroying plants or animals considered to be pests.

Petroleum: A naturally-occurring clear, green, or black liquid commonly refined into various types of fuels, including gasoline, kerosene, and diesel.

Pharmaceutical: A drug, medicine, or medication.

Phosphor: A material that gives off light when exposed to ultraviolet light or electricity.

Phthalates: A group of chemicals used to make plastics more flexible and harder to break.

Polonium-210: A naturally-occurring highly radioactive element.

Polychlorinated Biphenyls (PCBs): A group of man-made chemicals used mainly as coolants and lubricants in electrical equipment and motors because they don't burn easily and are good insulators.

Polychlorinated Dibenzofurans: Chemicals formed as by-products in the production and use of polychlorinated biphenyls and other compounds.

Polycyclic Aromatic Hydrocarbons (PAHs): A group of aromatic chemicals formed during the incomplete burning of garbage.

Preservative: A substance added to products to prevent decomposition.

Radioactive: A material that gives off radiation.

Radiation: A type of energy that comes from a source in the form of invisible waves or rays.

Rare Earth Metals: A set of natural but rarely-occurring elements found in the Earth's crust that are essential for the production of elecronic and other products.

Roundup: A chemical used to kill a wide variety of plants.

Sanitary Landfill: A site where waste is disposed of by burial and kept from contaminating the environment.

Shelltox: A chemical substance used to kill insects.

Styrene: A man-made chemical used extensively in the manufacture of plastics, rubber, and resins.

Styrofoam: A lightweight plastic foam with very good insulating properties used in a variety of products, including packaging materials, food containers, and drinking cups.

Sulfur Dioxide: A strong smelling, toxic gas formed as a result of burning petroleum.

Sulfuric Acid: A highly corrosive viscous acid with many uses, including lead-acid batteries, cleaners, oil refining, and fertilizer manufacturing.

Toxic: Capable of causing injury or death.

Triclosan: An ingredient added to many consumer products to reduce or prevent bacterial contamination.

Uranium: A very heavy, naturally-occurring radioactive metal.

Vinyl Chloride: A colorless flammable gas that evaporates very quickly—used to make polyvinyl chloride for plastic pipes, wire coatings, vehicle upholstery, and plastic kitchen ware.

Volatile: Evaporates readily at normal temperature and pressure.

Volatile Organic Compound: Chemical substances that contain carbon and evaporate easily at normal temperature.

Waste Transfer Station: A facility where garbage is unloaded from collection vehicles and processed for recycling, compaction, and reloading onto larger, long-distance vehicles for transport to a sanitary landfill.

Waste Pre-Treatment: A process or combination of processes by which waste is prepared before final disposal that includes separation into materials for recycling or reuse.

Western Corridor: An east-to-west stretch running across the middle of Belize.

References

Agency for Toxic Substances and Disease Registry (ATSDR). (1995). Toxicological profile for acetone. Atlanta, GA: U.S. Department of Health and Human Services, Public Health Service.

ibid. (1997) Toxicological profile for tetrachloroethylene.

ibid. (2007) Toxicological profile for lead.

ibid. Toxicological profile for arsenic.

ibid. Toxicological profile for benzene.

ibid. Toxicological profile for asbestos.

ibid. Toxicological profile for americium.

ibid. Toxicological profile for cadmium.

ibid. Toxicological profile for chlorine.

ibid. Toxicological profile for chloroform.

ibid. Toxicological profile for chromium.

ibid. Toxicological profile for DDT.

ibid. Toxicological profile for formaldehyde.

ibid. Toxicological profile for malathion.

ibid. Toxicological profile for methyl parathion.

ibid. Toxicological profile for Mercury.

ibid. Toxicological profile for phenol.

ibid. Toxicological profile for PCBs.

ibid. Toxicological profile for PAHs.

ibid. Toxicological profile for styrene.

ibid. Toxicological profile for vinyl chloride.

ibid. Toxicological profile for xylenes.

Alberta Environment. (2000). Fluorescent Lamp Stewardship Initiative. Retrieved August 21, 2014 <environment.gov.ab.ca/info/library/6344.pdf>.

Badreshia, S. (2002). Iodopropynyl butylcarbamate. *American Journal of Contact Dermatitis*, 13(2), 77–79.

Bakhiya, N. & Appel, K.E. (2010). Toxicity and carcinogenicity of furan in human diet. *Archives of Toxicology*, 84(7), 563–578.

Bandyopadhyay, A. & Basak, C. (2007). Studies on photocatalytic degradation of polystyrene. *Materials Science and Technology*, 23(3), 307–317.

Belize Legal Information Network. (2014). Solid Waste Management Authority Act (Chapter 224, Revised Edition 2003). Retrieved August 21, 2014 <belizelaw.org/web/lawadmin/index2.html>

ibid. Environmental Protection Act (Chapter 328, Revised Edition 2003).

ibid. Public Health Act (Chapter 40).

ibid. Belize City Council Act (Chapter 85).

ibid. Environmental Tax Act (Chapter 64:01).

ibid. Belize Port Authority Act (Chapter 233).

ibid. Water Industry Act (Chapter 222).

ibid. Pesticides Control Act (Chapter 216).

ibid. Returnable Containers Act (2009).

ibid. Scrap Metal Recyclers Regulation (2011).

ibid. Dangerous Goods Act.

Belize Ministry of Health. (2012). Mercury in fish from the Macal River, May 12 advisory. Retrieved August 21, 2014 <health.gov.bz/www/general-health/651-mercury-in-fish-from-the-macal-river-may-2012-advisory>.

Belize National Malaria Eradication Service. (2003). *Annual Quantity of DDT used in Belize for the Period 1985-1994.* Belmopan, Belize: Ministry of Health.

Belize Solid Waste Management Authority. (2011). *Waste Generation and Composition Study for the Western Corridor, Belize C.A.* 2056/OC-BL – Final Report.

Belize Solid Waste Management Authority. (2013). *Belize National Solid Waste Management Policy, Strategy and Plan – Inception Report.*

Belize Solid Waste Management Authority. (2014). Waste Transfer/Disposal. Retrieved August 22, 2014 <belizeswama.com/waste-transfer>.

Besis, A. & Samara, C. (2012). Polybrominated Diphenyl Ethers (PBDEs) in the Indoor and Outdoor Environments—A Review on Occurrence and Human Exposure." *Environmental Pollution*, 169, 217–229.

Beychok, M.R. (1987). A Data Base for Dioxin and Furan Emissions from Refuse Incinerators. *Atmospheric Environment*, 21(1), 29–36.

Biedermann, S., Tschudin, P., & Grob, K. (2010). Transfer of Bisphenol A from Thermal Printer Paper to the Skin. *Analytical and Bioanalytical Chemistry*, 398(1), 571–576.

Bonner, M.R., Coble, J., & Blair, A. et al. (2007). Malathion Exposure and the Incidence of Cancer in the Agricultural Health Study. *American Journal of Epidemiology*, 166(9), 1023–1034.

Center for Disease Control and Prevention (CDC). (2003). HIV and its Transmission. Retrieved August 21, 2014 <web.archive.org/web/20050204141148/http://www.cdc.gov/HIV/pubs/facts/transmission.htm>

Cherry, N., Moore, H., McNamee, R., Pacey, A., Burgess, G., Clyma, J., et al. (2008). Occupation and Male Infertility: Glycol Ethers and Other Exposures. *Occupational and Environmental Medicine*, 65(10), 708–714.

Cialdini, R.B., Raymond, R.R., & Kallgren, C.A. (1990). A Focus Theory of Normative Conduct: Recycling the Concept of Norms to Reduce Littering in Public Places. *Journal of Personality and Social Psychology*, 58(6), 1015-1026.

Cullinan, M.P., Palmer, J.E., Carle, A.D., West, M.J., & Seymour, G.J. (2012). Long Term Use of Triclosan Toothpaste and Thyroid Function. *Science of the Total Environment*, 416, 75–79.

Daughton, C.G., & Ternes, T.A. (1999). Pharmaceuticals and Personal Care Products in the Environment: Agents of Subtle Change? *Environmental Health Perspectives*, 107(6), 907-938.

Diaz, R.J., & Rosenberg, R. (2008). Spreading Dead Zones and Consequences for Marine Ecosystems. *Science*, 321(5891), 926–929.

EarthBasics. (2008). All about biodegradable plastic. Retrieved August 21, 2014 <earthbasics.com.au/page165bioplastic.html>.

Engler, R.E. (2012). The Complex Interaction between Marine Debris and Toxic Chemicals in the Ocean. *Environmental Science and Technology*, 46(22), 12302–12315.

Environmental Working Group (EWG). (2008). Comments for Public Meeting on "International Cooperation on Cosmetics Regulations (ICCR) Preparations". Retrieved August 21, 2014, <ewg.org/news/testimony-official-correspondence/comments-public-meeting-international-cooperation-cosmetics>.

Food and Drug Administration (FDA). (2014). Mercury Levels in Commercial Fish and Shellfish (1990-2010). Retrieved August 21, 2014, <fda.gov/Food/FoodborneIllnessContaminants/Metals/ucm115644.htm>.

Fukazawah, H.K., Hoshino, K., Shiozawa, T., Matsushita, H., & Terao, Y. (2001). Identification and Quantification of Chlorinated Bisphenol A in Wastewater from Wastepaper Recycling Plants. *Chemosphere*, 44(5), 973–979.

Golden R., Gandy, J., &Vollmer, G. (2005). A Review of the Endocrine Activity of Parabens and Implications for Potential Risks to Human Health. *Critical Reviews in Toxicology*, 35(5), 435–458.

Gregory, M.R. (2009). Environmental Implications of Plastic Debris in Marine Settings—Entanglement, Ingestion, Smothering, Hangers-On, Hitch-Hiking and Alien Invasions. *Philos Trans. R. Soc. Lond. B. Biol. Sci.*, 364(1526), 2013–2025.

Harrison, J., Leggett, R., Lloyd, D., Phipps, A., & Scott, B. (2007). Polonium-210 as a Poison. *Journal of Radiological Protection*, 27(1), 17–40.

Heudorf, U., Mersch-Sundermann, V., & Angerer, J. (2007). Phthalates: Toxicology and Exposure. *International Journal of Hygiene and Environmental Health,* 210 (5), 623–634.

Hollis, J.M., Lovas, F.J., Jewell, P.R., & Coudert, L.H. (2002). Interstellar Antifreeze: Ethylene Glycol. *The Astrophysical Journal*, 571(1), 59–62.

Kochukov, Y., Jeng, J. & Watson, S. (2009). Alkylphenol Xenoestrogens with Varying Carbon Chain Lengths Differentially and Potently Activate Signaling and Functional Responses in GH3/B6/F10 Somatomammotropes. *Environmental Health Perspectives*, 117(5), 723–730.

Kudo, N., & Kawashima, Y. (2003). Toxicity and Toxicokinetics of Perfluorooctanoic Acid in Humans and Animals. *Journal of Toxicological Sciences*, 28(2), 49–57.

Liem, A.K., Furst, P. & Rappe, C. (2000). Exposure of Populations to Dioxins and Related Compounds. *Food Addit. Contam.* 17, 241-259.

Liesivuori, J., & Savolainen, H. (1991). Methanol and Formic Acid Toxicity: Biochemical Mechanisms. *Pharmacology and Toxicology*, 69(3), 157–163.

Maia, M.F. & Moore, S.J. (2011). Plant-based Insect Repellents: A Review of Their Efficacy, Development and Testing. *Malaria Journal*, 10(1).

Mato, Y. (2001). Plastic Resin Pellets as a Transport Medium for Toxic Chemicals in the Marine Environment. *Environmental Science & Technology*, 35(2), 318-324

Mazzone, P. J. (2008). Analysis of Volatile Organic Compounds in the Exhaled Breath for the Diagnosis of Lung Cancer. *Journal of Thoracic Oncology*, 3(7), 774–780.

McField, M. (2005). *Analysis of Agrochemical Contamination of Marine Organisms from the Sapodilla Cayes.* Belize City, Belize: World Wildlife Fund.

Melnick, R.L., & Michael, C. (1995). Mechanistic Data Indicate that 1,3-Butadiene is a Human Carcinogen. *Carcinogenesis*, 16(2), 157–163.

Moore, C.G., & Phillips, C. (2011). *Plastic Ocean*. Penguin Group. ISBN 9781452601465.

Moore, C.J., et al. (2001). A Comparison of Plastic and Plankton in the North Pacific Central Gyre. *Marine Pollution Bulletin*, 42(12), 1297–1300.

Orloff, K., & Falk, H. (2003). An international perspective on hazardous waste practices. *International Journal of Hygiene and Environmental Health*, 206(4-5), 291-302.

Price, W., & Smith, E.D. (2006). Waste tire recycling: environmental benefits and commercial challenges. *International Journal of Environmental Technology and Management*, 6(3-4), 363-364.

Prockop, L.D, & Chichkova, R.I. (2007). Carbon Monoxide Intoxication: An Updated Review. *Journal of the Neurological Sciences*, 262(1–2), 122–130.

Ramanathan, V. (1975). Greenhouse Effect Due to Chlorofluorocarbons: Climatic Implications. *Science, New Series*, 19, 50–52.

Ray, A. (2008). Waste Management in Developing Asia: Can Trade and Cooperation Help? *The Journal of Environment & Development,* 17(1), 3-25.

Reinhardt, P.A., & Gordon, J.G. (1991). *Infectious and Medical Waste Management*. Chelsea, Michigan: Lewis Publishers.

Revkin, A.C. (1983). Paraquat: A Potent Weed Killer Is Killing People. *Science Digest*, 91 (6), 36–38.

Ryan, P.G. & Rice, N. (1996). The Free Shopping Bag Debate: Costs and Attitudes. *South African Journal of Science*, 92(4), 163-64.

Sakamato H., Matsuzawa, A., Itoh, R., & Tohyama, Y. (2000). Quantitative Analysis of Styrene Dimer and Trimers Migrated from Disposable Lunch Boxes. *J Food Hyg Soc Japan*, 41(3), 200–205.

Siegrist, H., Ternes, T.A., & Joss, A. (2004). Scrutinizing Pharmaceuticals and Personal Care Products in Wastewater Treatment. *Journal of Environmental Science & Technology*, 38, 392A-399A.

Snyder, S.A., Westerhoff, P., Yoon, Y., & Sedlack, D.L. (2003). Pharmaceuticals, Personal Care Products, and Endocrine Disruptors in Water: Implications for the Water Industry. *Environmental Engineering Science*, 20(5), 449-469.

Spivey, A. (2003). Plastic Bags—Prolific Problems. *Environmental Health Perspectives*, 111 (4), A208.

Sthiannopkao, S., & Wong, M.H. (2013). Handling E-Waste in Developed and Developing Countries: Initiatives, Practices, and Consequences. *Science of the Total Environment*, 463-464: 1147-53.

Storelli, M.M. (2000). Fish for Human Consumption: Risk of Contamination by Mercury. *Food Additives and Contaminants*, 1007–1011.

Thompson, R.C., Olsen, Y., Mitchell, R.P., Davis, A., Rowland, S.J., John, A.W., et al. (2004). Lost at Sea: Where is All the Plastic? *Science*, 304(5672), 838.

Tränkner, A. (1992). Use of Agricultural and Municipal Organic Wastes to Develop Suppression to Plant Pathogens. In E.C. Tjamos, G.C. Papavizas, and R.J. Cook (Eds.), *Biological Control of Plant Diseases*. New York: Plenum Press.

U.S. Energy Information Administration (USEIA). (2012). More Recycling Raises Average Energy Content of Waste Used to Generate Electricity. Retrieved August 22, 2014, <eia.gov/todayinenergy/detail.cfm?id=8010>.

United States Department of Labour. Occupational Safety & Health Administration (OSHA). (2014). *Methylene Chloride.* Retrieved August 22, 2014, <osha.gov/pls/oshaweb/owadisp.show_document?p_table=STANDARDS&p_id=10094>.

United States Environmental Protection Agency (USEPA). (2003). *You Dump It, You Drink It: Recycle Used Motor Oil.* Retrieved August 21, 2014, <epa.gov/epawaste/conserve/materials/usedoil/campgn/en-dumpbr.pdf>.

United States Environmental Protection Agency (USEPA). (2010). *Refrigerant Safety.* Retrieved August 21, 2014, <epa.gov/ozone/snap/refrigerants/safety.html>

United States Environmental Protection Agency (USEPA). (2012). *Health and Environmental Concern/Tires.* Retrieved November 2, 2013, <epa.gov/wastes/conserve/materials/tires/health.htm>.

United States Environmental Protection Agency (USEPA). (2013). *Benefits of Recycling.* Retrieved January 12, 2014, <epa.gov/recycle/recycling-basics>.

United States Environmental Protection Agency (USEPA). (2006). *EPA Improves Standards for Recycling of Cathode Ray Tubes.* Retrieved August 22, 2014, <yosemite.epa.gov/opa/admpress.nsf/2006+press+releases/373c5b81f7dd8111852571b00062cd2d?opendocument>.

United States Environmental Protection Agency (USEPA). (2014). *Household Hazardous Waste.* Retrieved August 22, 2014, <epa.gov/epawaste/conserve/materials/hhw.htm>.

United States Environmental Protection Agency (USEPA). (2014). *Sulfur Dioxide.* Retrieved August 22, 2014, <epa.gov/airquality/sulfurdioxide>.

United States Environmental Protection Agency (USEPA). (2014). *Uranium.* Retrieved August 22, 2014, <epa.gov/radiation/radionuclides/uranium.html>.

Veenstra, G. & Webb, C. (May 2009). Human Health Risk Assessment of Long Chain Alcohols. *Ecotoxicology and Environmental Safety*, 72(4), 1016-1030.

Von Goetz, N., Wormuth, M., Scheringer, M., & Hungerbühler, K. (2010). Bisphenol A: How the Most Relevant Exposure Sources Contribute to Total Consumer Exposure. *Risk Analysis*, 30 (3), 473–874.

Williams, G.M., Kroes, R., & Munro, I.C. (2000). Safety Evaluation and Risk Assessment of the Herbicide Roundup and its Active Ingredient, Glyphosate, for Humans. *Regulatory Toxicology and Pharmacology*, 31(2 Pt 1), 117–165.

World Health Organization (WHO). (2014). Dioxins and Their Effects on Human Health. Retrieved August 21, 2014, <who.int/mediacentre/factsheets/fs225/en>.

World Health Organization (WHO). (2014). *Hepatitis.* Retrieved August 21, 2014, <who.int/topics/hepatitis/en>.

World Health Organization (WHO). (2014). *Tuberculosis*. Retrieved August 21, 2014, <who.int/mediacentre/factsheets/fs104/en>.

Yabannavar, A.V. & Bartha, R. (1994). Methods for Assessment of Biodegradability of Plastic Films in Soil. *Applied and Environmental Microbiology*, 60, 3608-3614.

Yilmaz, D. (2011). In the Context of Visual Pollution: Effects to Trabzon City Center Silhoutte. *The Asian Social Science Journal*, 7(5), 99.